Alternative Energy

WATER POWER

Louise Kay Stewart

First published in Great Britain in 2023 by Hodder & Stoughton

Editor: Elise Short
Design: Paul Cherrill
Illustration: DGPH Studio

The text in this book was previously published
in the series *Let's Discuss Energy Resources* and has been updated.

HB ISBN: 978 1 5263 2527 3
PB ISBN: 978 1 5263 2530 3

Printed and bound in China

FSC
www.fsc.org
MIX
Paper from
responsible sources
FSC® C104740

Wayland, an imprint of
Hachette Children's Group
Part of Hodder & Stoughton
Carmelite House
50 Victoria Embankment
London EC4Y 0DZ

An Hachette UK Company
www.hachette.co.uk
www.hachettechildrens.co.uk

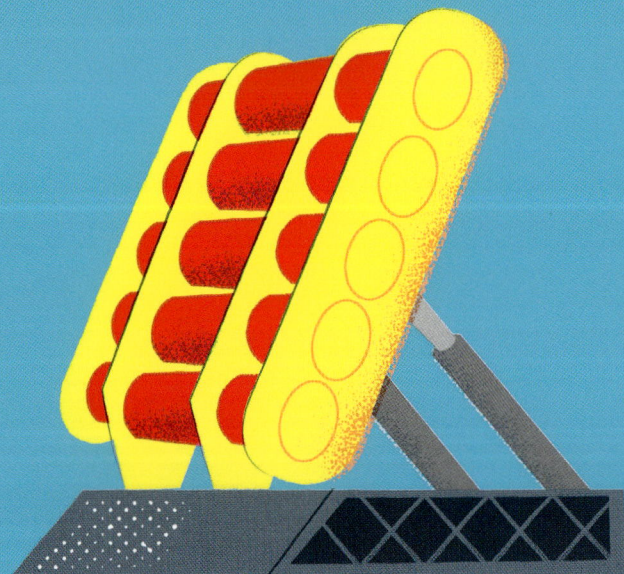

CONTENTS

Water power as an energy resource 4

Hydroelectric power 6

Hydroelectric dams 8

Hydroelectric and the environment 10

Impacts of dams on people 12

Hydroelectric power without dams 14

Tidal power 16

Wave power 18

Problems with tidal and wave power 20

Demand and supply of hydropower 22

Increasing water power 24

New water power technology 26

The future of water power 28

Think further 30

Glossary 31

Index 32

WATER POWER AS AN ENERGY RESOURCE

Water is an energy resource that people have used for centuries to help them to do their work. In the past, the power of moving water turned giant wheels that ground wheat or made machines work. Today, most water power is used to generate electricity.

Growing demand

We are using far more electricity than in the past. Most electricity is generated by burning fossil fuels, such as coal and gas. Fossil fuels formed millions of years ago from ancient living things. They are known as non-renewable energy resources and we are running out of them.

Fossil fuel problems

There's another problem with fossil fuels. Burning fossil fuels in power stations releases greenhouse gases into the air. Greenhouse gases build up in Earth's atmosphere and trap heat from the Sun. This is causing climate change. Climate change is making places much hotter, drier or wetter than they once were and causing extreme weather, such as floods and droughts.

Renewable energy sources

Alternatives to fossil fuel power are renewable energy resources, such as sunlight, wind and moving water. Most water power is hydroelectric power, made using fresh water flowing in rivers or from reservoirs. Tidal power uses the movement of ocean tides, while wave power uses the movement of ocean waves to generate electricity.

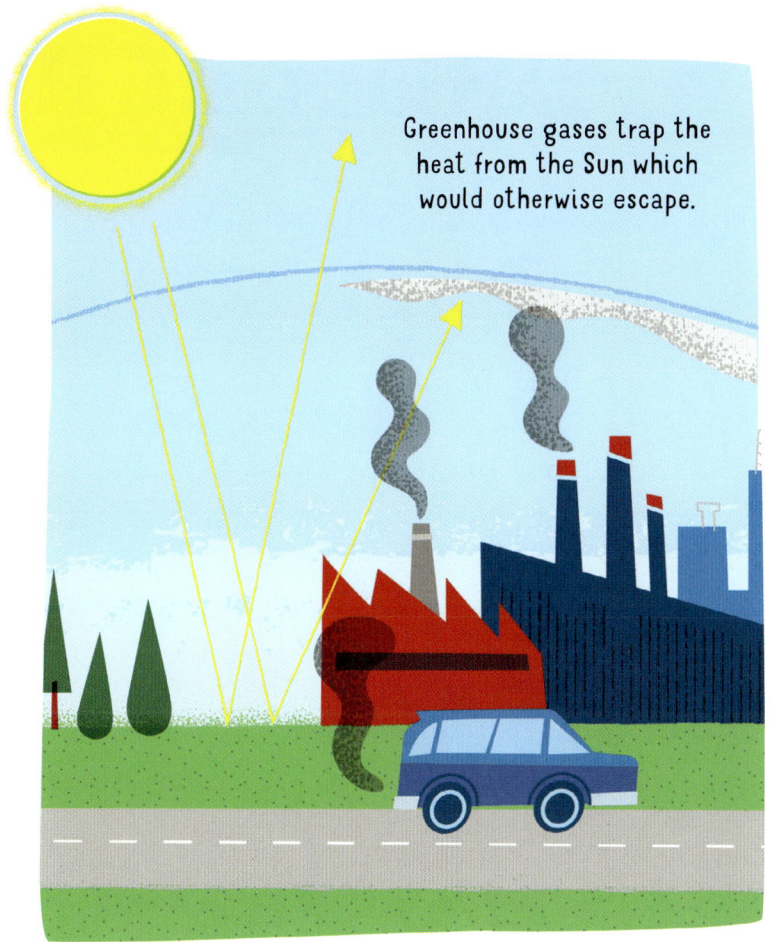

Greenhouse gases trap the heat from the Sun which would otherwise escape.

Power companies around the world control and use the flow of water to generate electricity.

HYDROELECTRIC POWER

Moving water contains a lot of energy. This is known as movement or kinetic energy. Hydroelectric power plants turn the energy from moving water into electrical energy or power.

Moving water

Water in a river moves as it flows downhill. The water flows more strongly where slopes are steeper and where there is more water. So, there is more force in a wide, deep river than in a shallow, narrow river moving down the same slope.

Rivers and reservoirs

Power companies mostly get the moving water for hydroelectric power from reservoirs. Reservoirs are deep, wide human-made lakes that trap river water. These create a large quantity, or head, of water. When the large amount of water in a reservoir is released through gates and down narrow tubes towards a turbine, it has powerful kinetic energy.

A dam is a structure built across a river or stream to hold back water and create a reservoir.

Turbine to electricity

- Moving water pushes against a turbine, which looks a bit like a propeller.
- The turbine's angled blades convert the strong push of water into a spinning movement.
- The turbine converts kinetic into mechanical energy in a shaft that spins a generator.
- This machine generates electricity by rotating coils of wire within a circle of magnets.

How hydroelectric power works

Electrical energy from generator

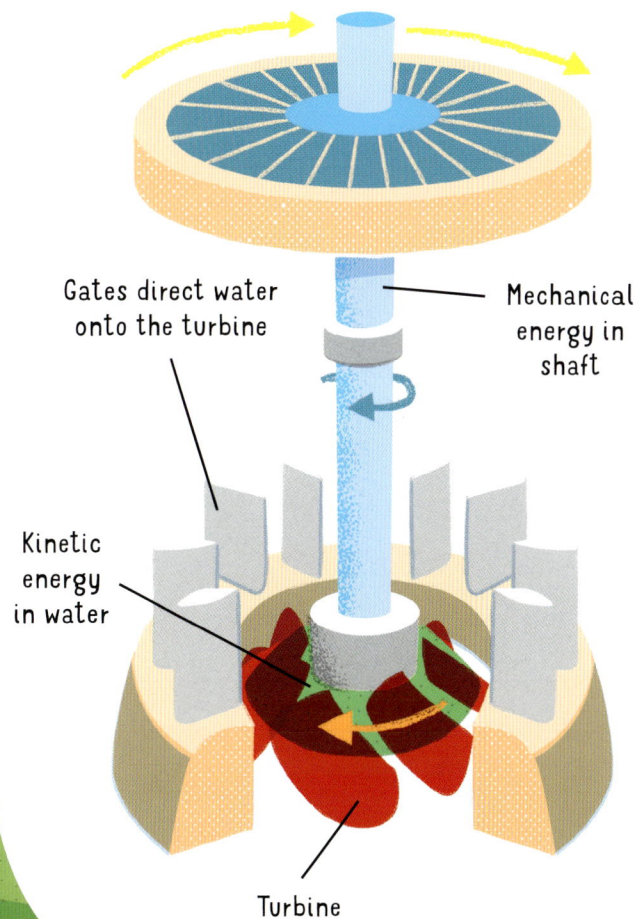

Gates direct water onto the turbine

Mechanical energy in shaft

Kinetic energy in water

Turbine

HYDROELECTRIC DAMS

Most hydroelectric reservoirs are made by building dams across rivers. The turbines and generators are built into the dam structure to form water power stations, or plants.

Building a dam

Engineers design hydroelectric dam walls to be strongest and thickest at the base. This is because the pressure or push of water is greatest at the bottom, due to the weight of water above. Building dams means constructing other kinds of infrastructure too, such as a network of cables and pylons to take the electricity away, and roads for workers to access the power station.

How a hydroelectric dam works

1. Sluice gates open or close to control water flow.

2. Water moves through a filter to remove debris that could damage the turbine.

3. Water rushes down a tunnel called a penstock.

The best location for hydroelectric dams is between rocky banks of powerful rivers that can refill reservoirs as water is used.

Dam locations

Many of the world's larger rivers, such as the Mekong in Vietnam and the Nile in Egypt, have hydroelectric dams. Countries with smaller rivers still build hydroelectric dams, they just need more of them. Norway has over 1,000 hydroelectric dams that together make the majority of the country's electricity.

5. A generator spins above each turbine to generate electricity.

6. Electricity is carried away through power lines.

4. It hits large turbines.

7. Water flowing out through the turbine is usually carried away in the river.

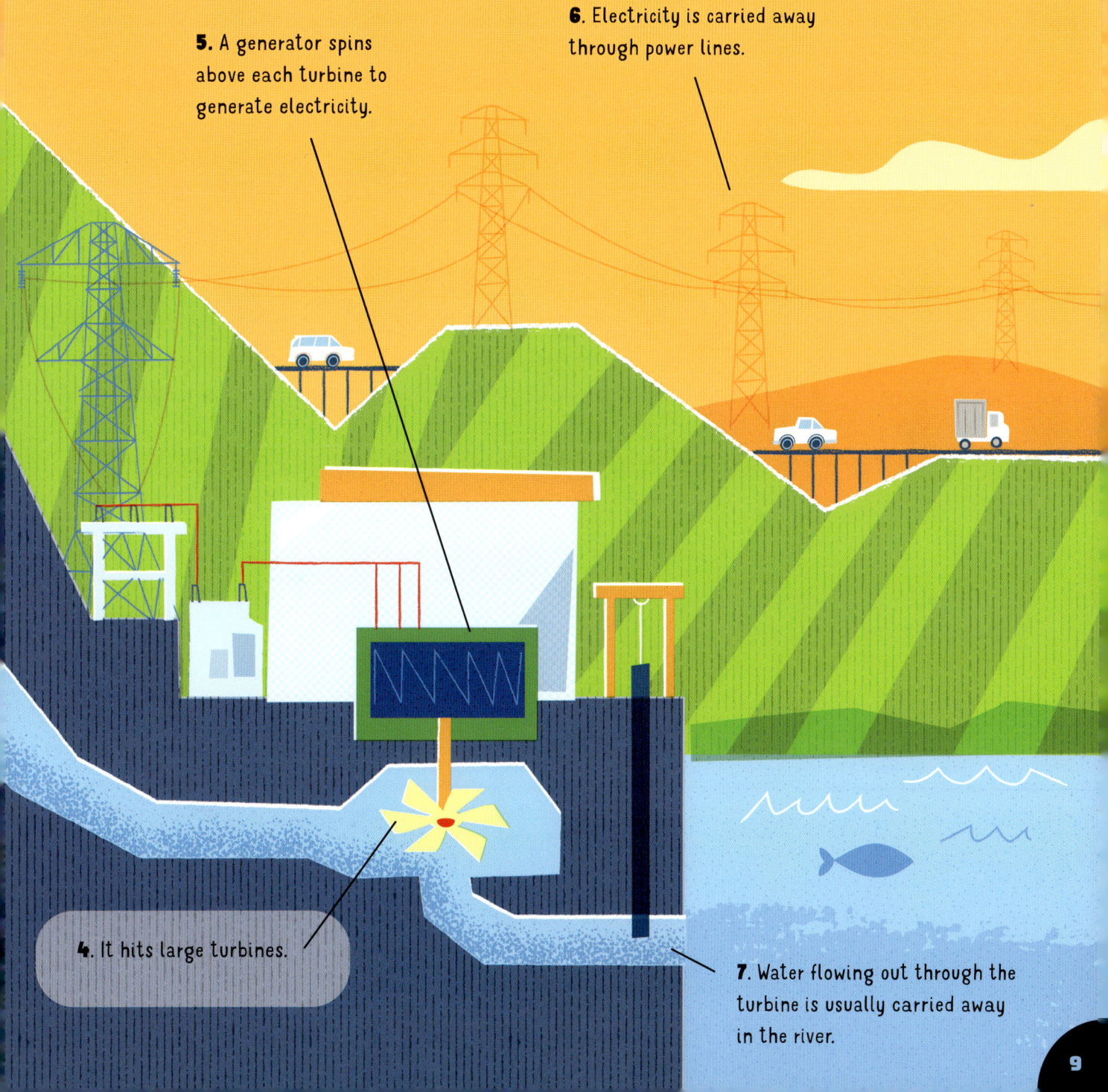

HYDROELECTRIC AND THE ENVIRONMENT

Hydroelectric dams block rivers to trap water in lakes and form reservoirs. Dam-building has many different impacts on rivers, land and the atmosphere.

Dams change environments

To create hydroelectric reservoirs, valleys, areas of farmland, woodland and other environments are flooded. Hydroelectric power stations are a major cause of deforestation, especially in tropical rainforest areas with big rivers. Another problem is that, when plants are buried underwater they rot, and as they do so they release a greenhouse gas called methane.

Water Use

Power companies release large amounts of reservoir water through dams when they need to make electricity. To create a powerful flow, they may hold back water for long periods of time that would have flowed downstream. Lakes and wetlands downstream get less water and may dry up.

Dead trees are all that remains of this forest, which was flooded to create a reservoir behind a hydroelectric dam.

Effects on wildlife

When rivers are blocked for reservoirs, some animals find it difficult to survive. Dams may block the movement of river animals, such as fish. For example, one reason the Atlantic salmon is endangered in the US is due to dams blocking the fish's route to breeding places.

The problem of reservoirs holding back water from downstream is made worse if people use reservoir water for other reasons, such as the irrigation of farmland or for drinking water.

IMPACTS OF DAMS ON PEOPLE

Building hydroelectric power plants can bring great benefits to an area. New industries move in to use the electricity, and they create jobs and wealth. However, new dams and reservoirs can cause problems for people, too.

Moving people

Land flooded to create a reservoir may have villages and farmland on it. So, large numbers of people may be displaced or forced to leave their homes in order for it to be built. When people are displaced, they often have no choice in where they are moved to. Some people end up in crowded towns where it is hard for them to find work.

Losing livelihoods

People downstream of dams may also end up being displaced. Areas of the river downstream that become slower and emptier are often unsuitable for fishing or for transporting goods. People in these industries then have to search for work elsewhere. There may be further displacement and also injuries or even deaths if dams burst and flood land downstream.

An estimated 80 million people have been displaced by dam projects worldwide in the last hundred years.

Illnesses caused by reservoirs

Areas of still water, such as reservoirs, are ideal places for insects, such as mosquitoes, to breed. In tropical countries, mosquitoes spread a disease called malaria that kills millions of people. There is also the danger that harmful chemicals can wash from buildings and factories that were flooded to make new reservoirs. This can make reservoir water poisonous.

A village flooded by the reservoir

Some experts say that the weight of water in big reservoirs can cause earthquakes, too.

HOMES NOT DAMS!

NO DAMS

NO DAMS

SAVE OUR HOMES!

NO D.

HOMES NOT DAMS!

SAVE OUR HOMES!

HYDROELECTRIC POWER WITHOUT DAMS

Around the world, small systems of turbines in rivers or other water channels supply the small amounts of electricity needed by individual families, businesses or communities. This kind of small-scale electricity production is known as microgeneration.

Why microgeneration?

People living in remote places often use microgeneration because they have no mains electricity. More than 1.5 billion people worldwide do not have mains electricity, because there is no grid (network of cables and pylons) to carry electricity from power stations to their homes and buildings..

Penstock

Powerhouse containing the turbine

Where microgeneration works

People can use microgeneration if there is a supply of fast-moving water nearby to operate turbines. Most hydroelectric microgeneration happens in mountains, where lots of water melts into streams and small rivers and flows down steep slopes.

Hydroelectric microgeneration supplies electricity to people who might otherwise have none.

How microgeneration works

The simplest microgeneration system has a turbine with a generator fixed in a steep river or stream. Other systems have a pool next to a river or waterfall with a strong flow of water. The pool is connected to a turbine lower down the slope by a long pipe or penstock.

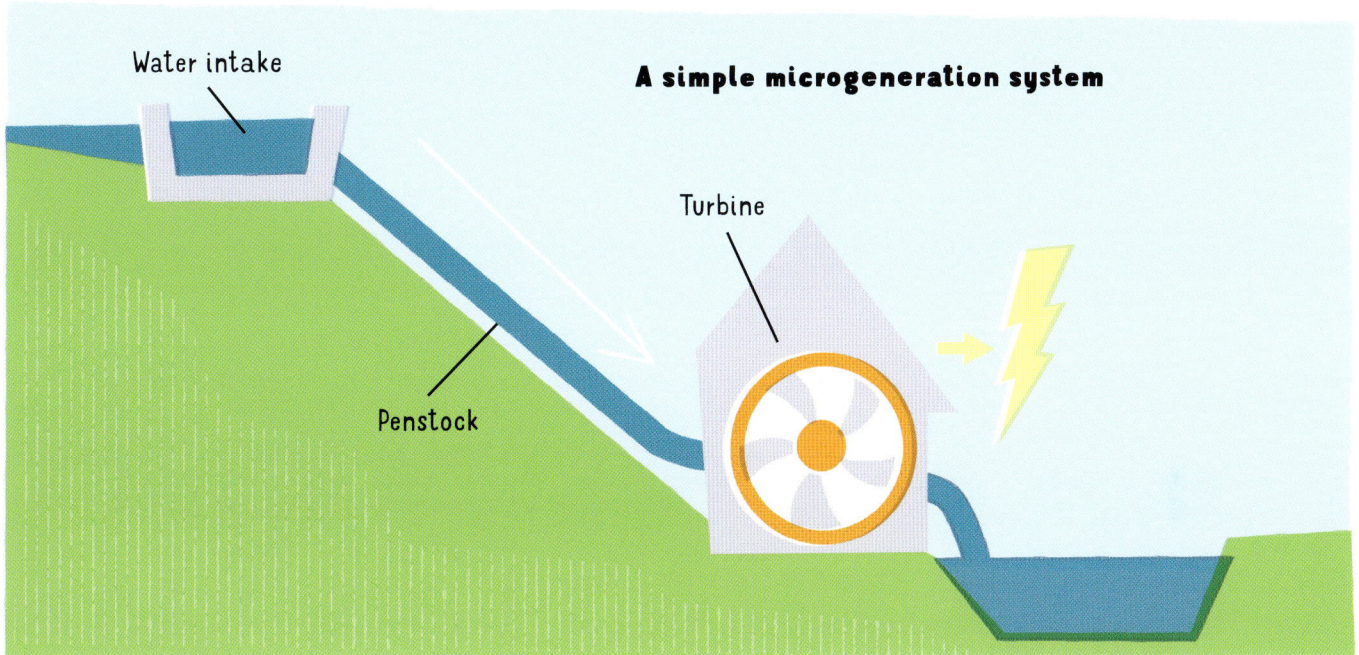

Water intake

A simple microgeneration system

Turbine

Penstock

Cheap and cheerful

Big hydroelectric systems that generate electricity for large numbers of people are built using expensive technology. Many microgeneration systems are made by local communities using cheaper, local materials, such as bamboo pipes or recycled metal.

Turbines used for microgeneration are driven by small amounts of fast-moving water.

TIDAL POWER

Tidal power systems use seawater to generate electricity. They harness the daily movement of tides towards and away from coastal land, or fast-moving currents in underwater channels.

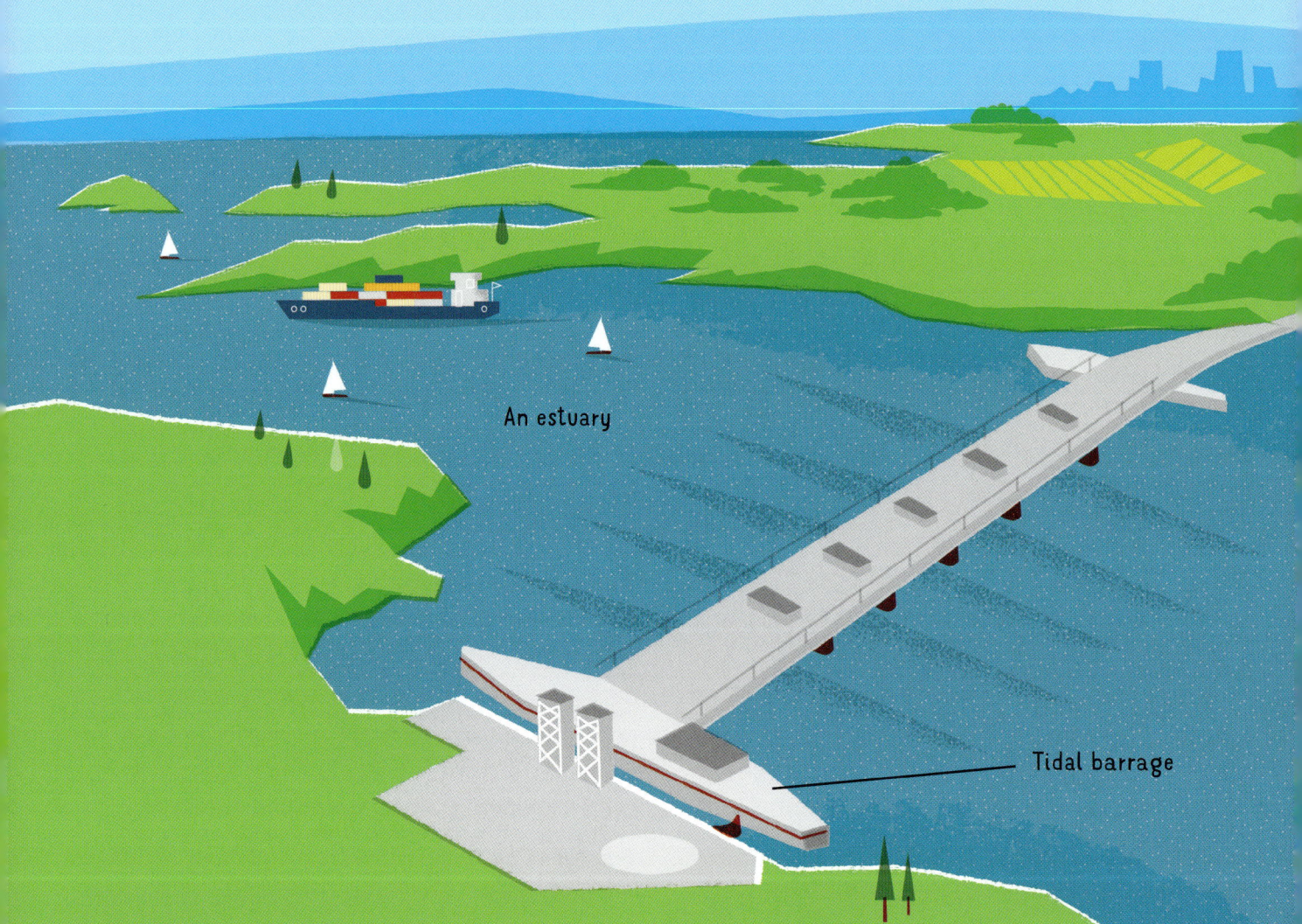

An estuary

Tidal barrage

Tidal barrages

Tidal barrages are like wide, low dams built across shallow bays and estuaries. These coastal areas are good places for barrages because their shape narrows towards land. They funnel tidal water moving in, so the water level gets higher than it would on a straight area of coast.

How tidal barrages work

When the tide comes in, the barrage blocks the flow of seawater towards land. The water is trapped on the sea side of the barrage. At high tide, sluice gates open and seawater flows past turbines inside the barrage. These turn generators that make electricity. When the tide starts to go out, the sluice gates are closed. Water is trapped on the land side. At low tide, this water is released to generate electricity.

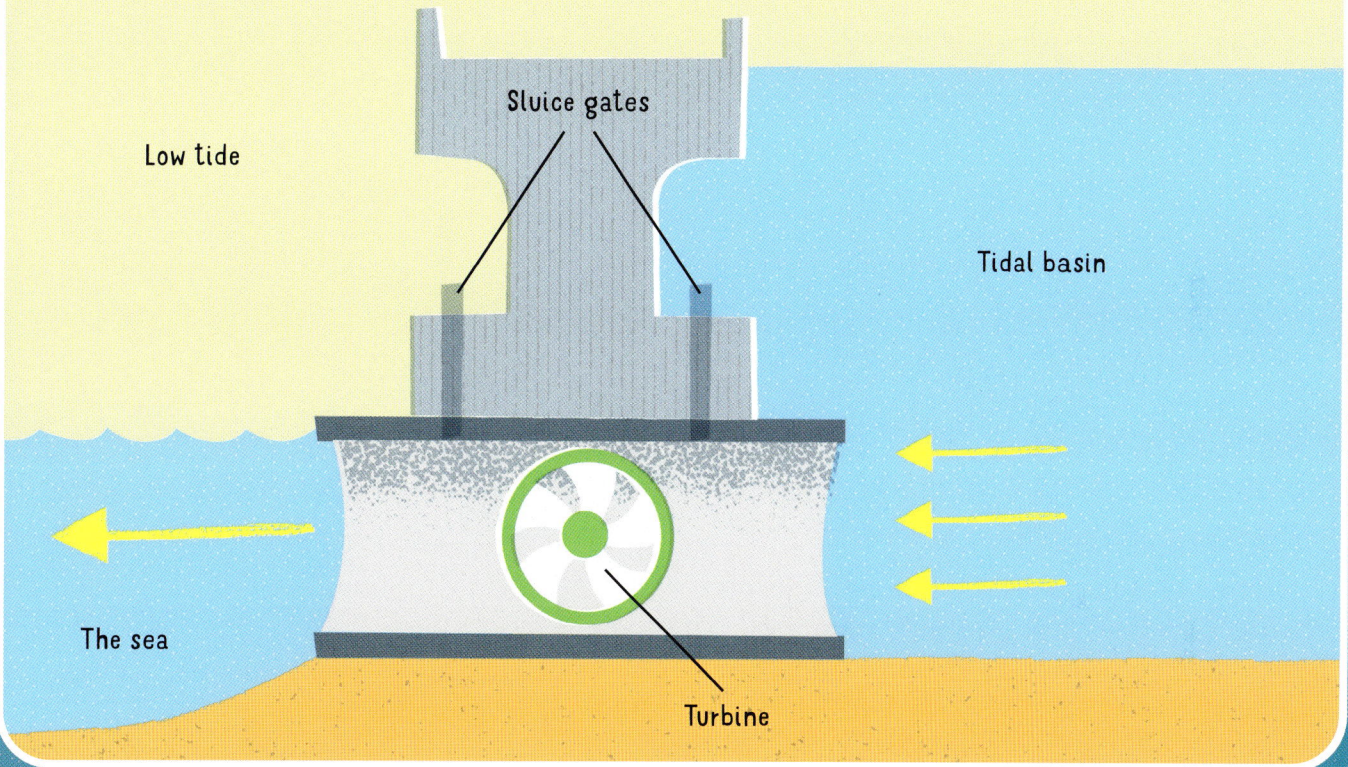

Low tide

Sluice gates

Tidal basin

The sea

Turbine

Underwater turbines

Some tidal power is made using underwater turbines. As tides come in and go out, water currents turn the turbine blades. Power companies usually install several turbines which they attach individually to the seafloor or build into underwater walls that they call reefs.

Wide underwater turbines capture the kinetic energy in underwater currents.

WAVE POWER

Wind blows against the surface of the sea and creates waves that are packed with kinetic energy. Wave power systems capture wave energy to turn a generator that generates electricity. The electricity travels through underwater cables to land. There are a lot of different types of wave power technology.

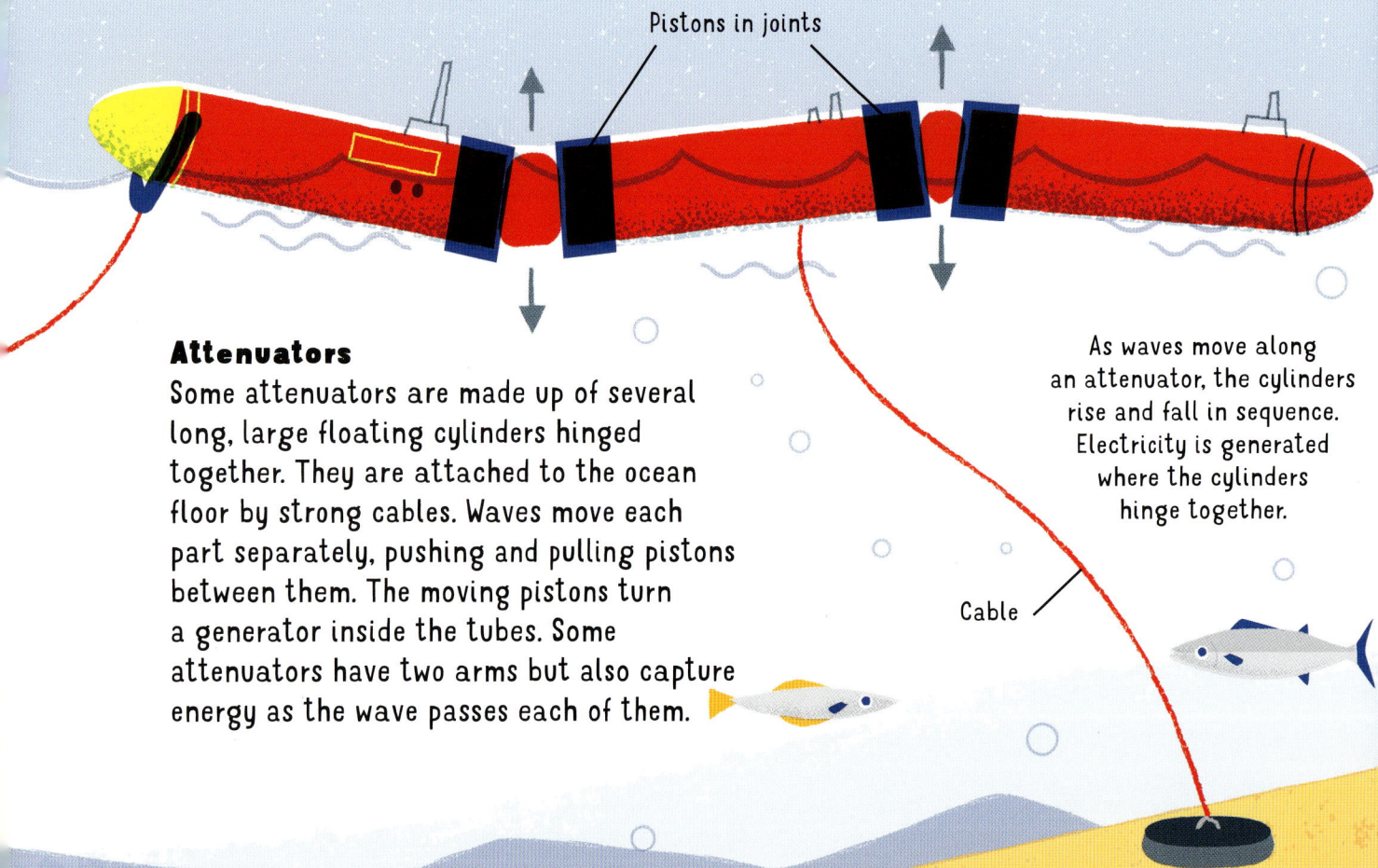

Pistons in joints

Attenuators

Some attenuators are made up of several long, large floating cylinders hinged together. They are attached to the ocean floor by strong cables. Waves move each part separately, pushing and pulling pistons between them. The moving pistons turn a generator inside the tubes. Some attenuators have two arms but also capture energy as the wave passes each of them.

As waves move along an attenuator, the cylinders rise and fall in sequence. Electricity is generated where the cylinders hinge together.

Cable

Oscillators

The simplest kind of oscillator has a part that looks like a giant arm that is fixed to a base unit on the ocean floor. Waves move the oscillator back and forth like a pendulum. This movement turns a generator on the base.

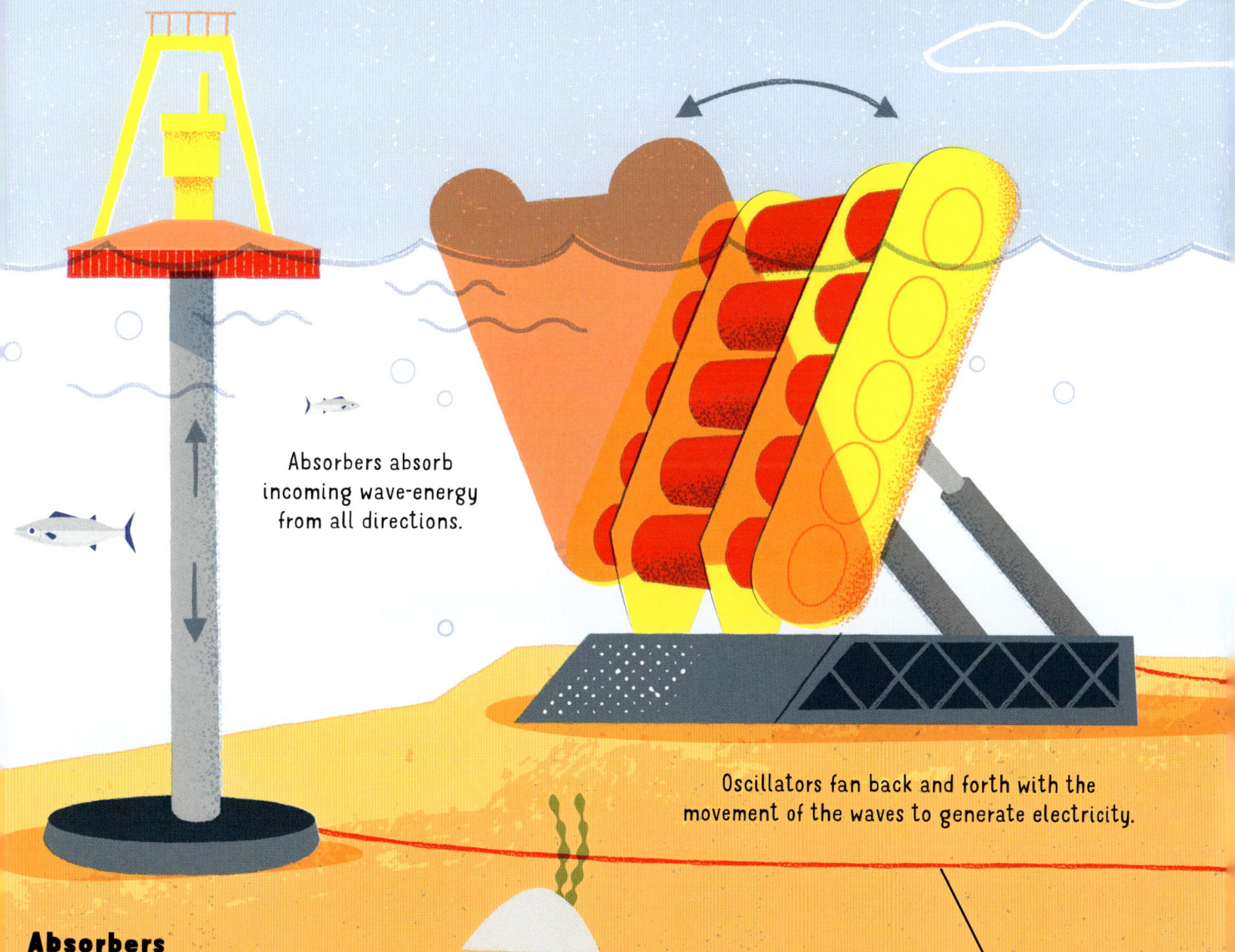

Absorbers absorb incoming wave-energy from all directions.

Oscillators fan back and forth with the movement of the waves to generate electricity.

Underwater cable carrying generated electricity

Absorbers

Absorbers have round floating top parts rather like buoys. These are anchored to the sea bed. The floating buoys rise and fall as waves pass by. This pulls pistons up and down and these turn a generator.

PROBLEMS WITH TIDAL AND WAVE POWER

Tidal and wave power are useful and important sources of renewable energy, but they do have some drawbacks. Both tidal and wave power can have some negative impacts on the environment.

Building tidal power plants

Tidal power plants can only be built in places where there are strong tides, meaning their locations are limited. Where they are built, they can change the structure of a coast and cause erosion of sand dunes. Dunes help to prevent flooding in many coastal areas. Tidal power plants can also prevent access to rivers and cut off shipping routes.

Wave power systems can only work when there are waves. They won't work on a calm day when there is no wind or waves.

Barrage impacts on wildlife

Tidal barrages destroy the habitat of estuary species, including wading birds. For example, avocets are wading birds that need coastal mud to find the food they need to survive. Spinning blades can kill marine wildlife.

An avocet

Turbines can obstruct shipping lanes.

Wave machine worries

It has been expensive to make wave machines big enough to produce large amounts of electricity. These machines can also change the habitat of animals, such as crabs and starfish, and spinning blades cause noise that disturbs the sea life around them. There is also a danger of toxic chemicals that are used on wave energy platforms spilling and polluting the water near them.

DEMAND AND SUPPLY OF HYDROPOWER

Hydropower stations are more suitable for some places than others and the amount of electricity they can supply also varies.

In many places with adequate rainfall, rivers flow day and night all year round, which allows hydropower stations to keep generating electricity.

Power price

Hydroelectric stations can supply electricity all year round in places where rivers flow constantly. Hydroelectricity is also cheaper than electricity from most other sources. That's because once a dam has been built and the sturdy, simple equipment has been installed, the flowing water (its energy source) is free.

Supply problems

In some places, hydroelectricity cannot meet demand, even if there are many hydroelectric stations. For example, in very hot places, such as Kenya, rivers may dry up. In very cold places, such as Canada, rivers may freeze over. Then power companies must generate electricity using other energy resources to meet demand.

Supply solutions

Some countries share hydropower, or buy it from each other. Norway produces almost all of its electricity from hydropower. Norway is connected to a Scandinavian grid network, so Sweden, Denmark and Finland can easily buy hydropower from Norway when that country has a plentiful supply.

Norway makes more hydroelectricity than it can use, so it is able to sell some to other countries.

INCREASING WATER POWER

Many countries around the world have agreed targets to tackle climate change by cutting the amount of greenhouse gases they produce. The problem is that they still need to supply enough electricity for their people and industries. One way to meet targets is to increase the amount of renewable energy, because this produces few or no greenhouse gases.

High Costs

The problem is that the cost of building new water power plants is high. Constructing these plants is difficult as turbines often need to be built in locations where there are high winds and giant waves. Once set up, the costs are low, but governments have to find ways to encourage companies to invest in renewable energy projects, such as hydropower.

All power technologies have some environmental problems, but it is better to subsidise a renewable rather than a non-renewable energy resource.

Paying for power

One way governments encourage renewable power is to pay more for renewable electricity than for fossil fuel power. Paying extra to make something cheaper so more people buy it is called subsidising. A lot of people think this is fair because fossil fuel companies are already subsidised in a way, because they don't pay for the effects of the climate change they are causing.

Building a dam is a massive, complex and expensive engineering project.

NEW WATER POWER TECHNOLOGY

For water power to reach its full potential, new technologies are needed to make wave, tidal and hydropower much more widely used than they are today.

More efficiency

Modern hydro turbine generators are getting more and more efficient. They can convert over 90 per cent of the energy in the available water into electricity. This is more efficient than any other form of electricity generation.

More micro-hydro

New and more efficient micro-hydropower systems are being created. They can use the energy in even small streams and waterways to drive a generator. Micro-hydropower systems can run on streams so small that they only supply enough electricity for one house.

Ocean heat to electricity

Ocean thermal energy conversion (OTEC) is a method of using the temperature difference between cold, deep seawater and warmer surface seawater to make electricity. OTEC systems pipe in warm and cold seawater and run them through water condensers and evaporators to spin turbines that generate electricity.

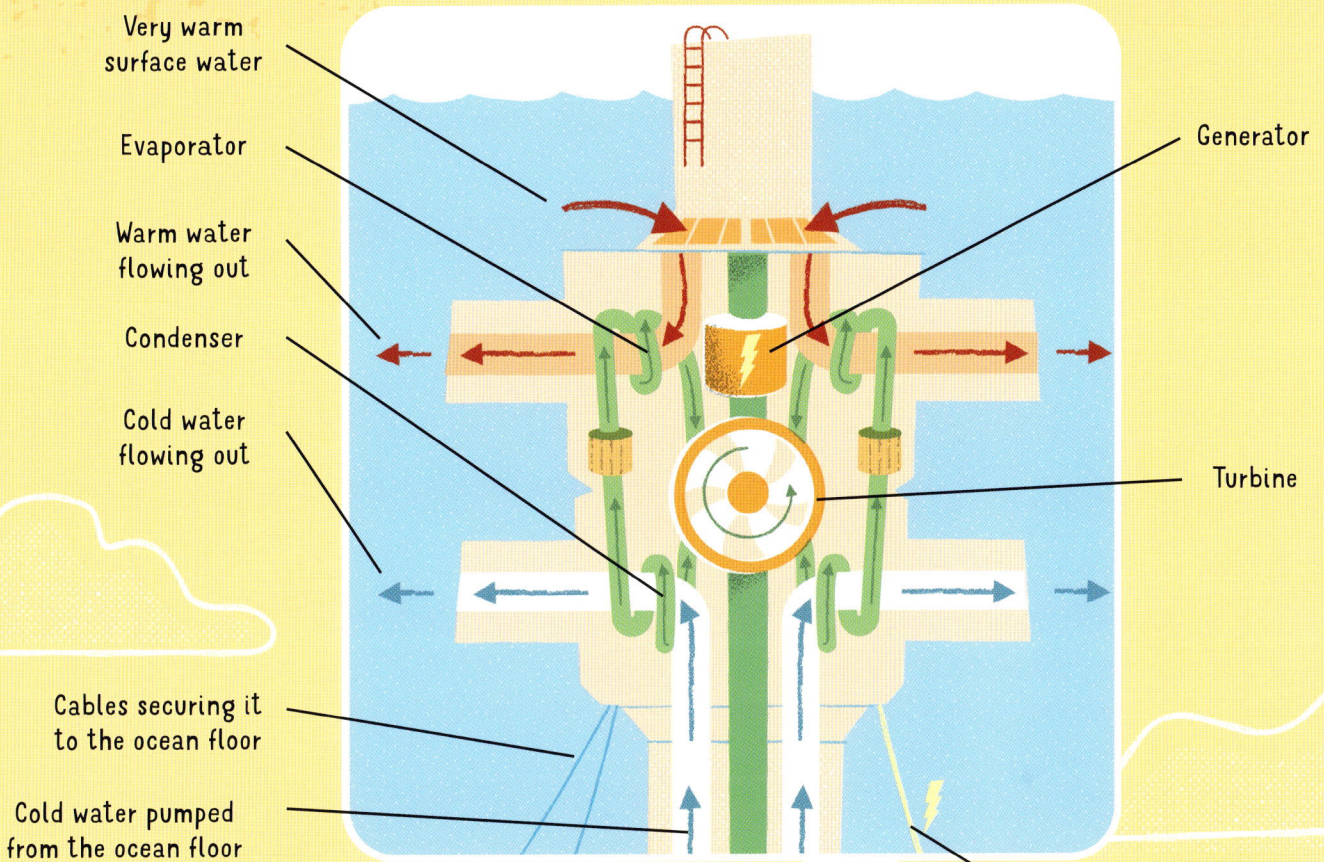

Very warm surface water

Evaporator

Warm water flowing out

Condenser

Cold water flowing out

Generator

Turbine

Cables securing it to the ocean floor

Cold water pumped from the ocean floor

Power cable taking electricity to shore

Ocean thermal energy systems on offshore floating structures use the temperature difference between surface and deep seawater to generate electricity. OTEC works best in tropical seas.

THE FUTURE OF WATER POWER

Water power is renewable, uses no fuel, and when up and running creates no greenhouse gases. Currently, water power produces almost 20 per cent of the world's electricity and over 90 per cent of the world's renewable power. How will water power develop in the future?

Hydroelectric power in the future

Hydroelectric power has the potential to supply half of the world's electricity needs. One way to generate more is to convert existing dams, which were built to store water for people to use in homes, business and farms. Some countries are already in the process of converting these into hydroelectric dams.

Coastal countries around the world are considering tidal current turbines as part of their energy mix.

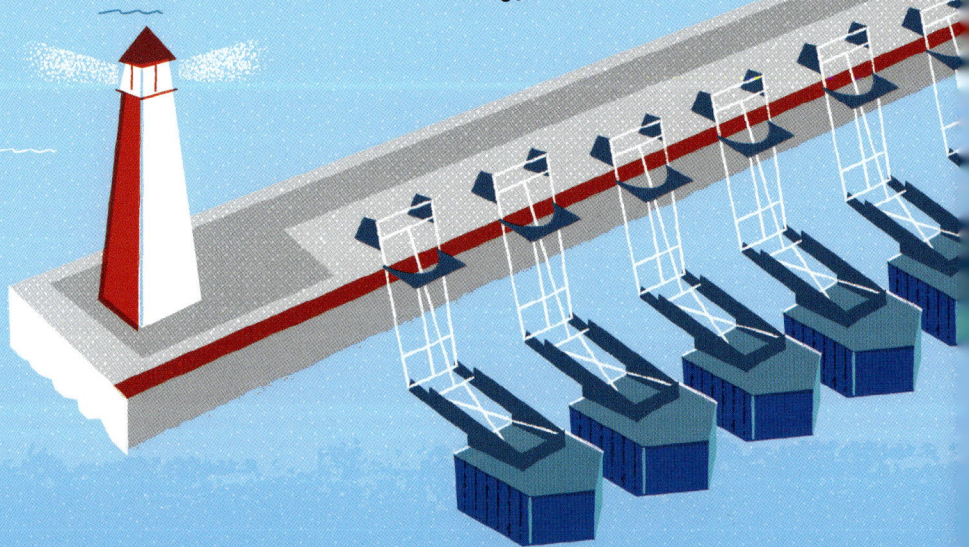

More microgeneration

Most new hydroelectric power systems built in the future are more likely to be in the form of microgeneration. Micro-hydro systems are cheaper to build than big hydroelectric projects, and do not carry the same risk of environmental problems.

Wave and tidal power

The main increase in wave and tidal power will come in places where there are powerful waves and high tides, such as off northwestern Europe and North America, southern Australasia, South America and Africa. The hope is that the technology to harness water power will also get cheaper if it is used more widely.

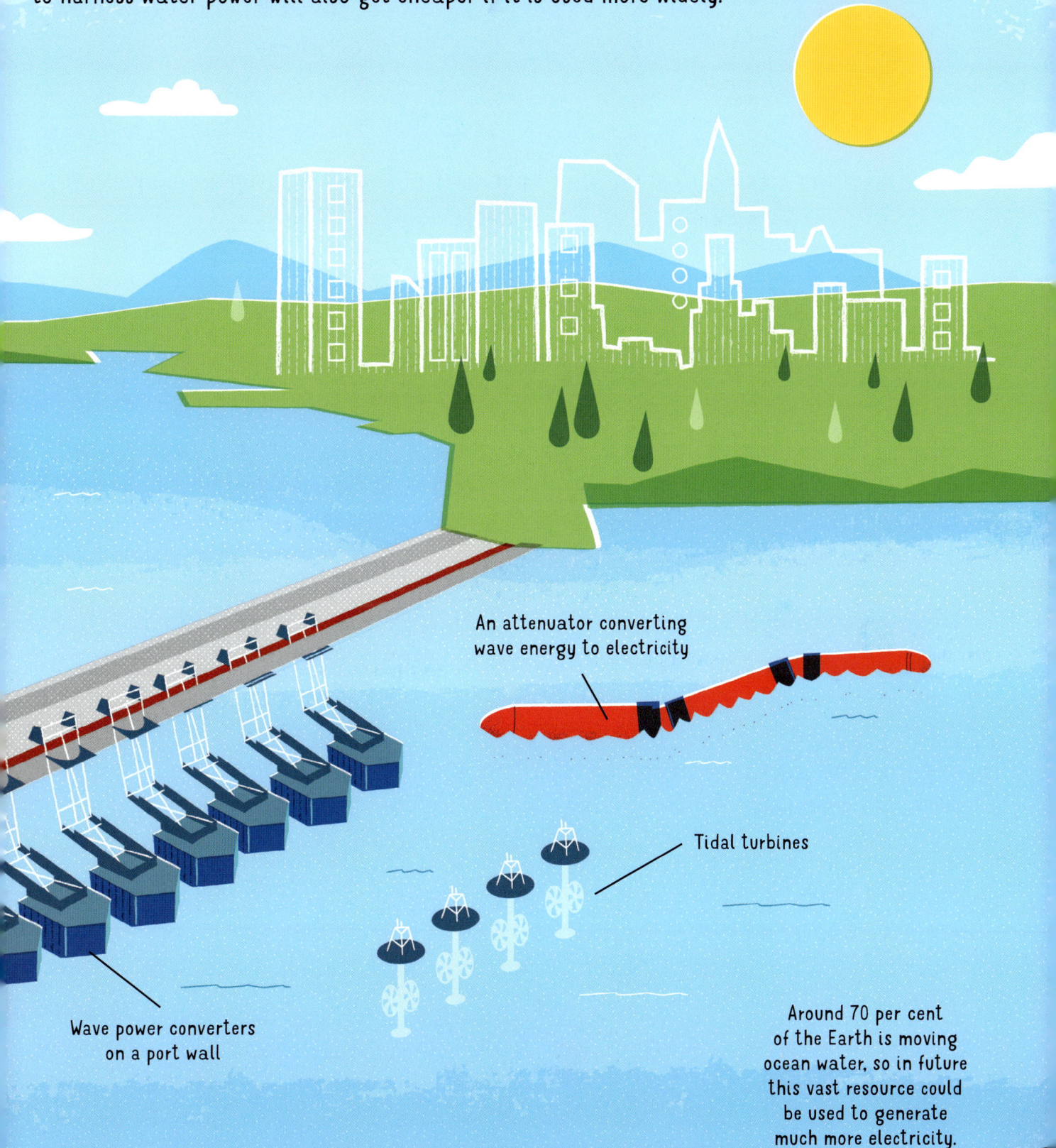

An attenuator converting wave energy to electricity

Tidal turbines

Wave power converters on a port wall

Around 70 per cent of the Earth is moving ocean water, so in future this vast resource could be used to generate much more electricity.

THINK FURTHER

Water power is a hot topic, with people often disagreeing about the pros and cons of using it to generate electricity. What do you think?

For and against

Hydroelectric power stations are expensive to build.
But ...
Once they are up and running, there is no fuel to buy so they are cheaper than fossil fuel plants.

When valleys are flooded to create hydroelectric reservoirs, trees and land are destroyed.
But ...
Hydroelectric dams burn no fuel, and their generators release no polluting gases that would cause environmental problems.

Coal, oil and gas are non-renewable resources that cost money and will one day run out.
But ...
Water power is renewable: tides rise and fall every day and there is always water in big rivers and seas.

Which of these statements do you agree with? Why? You could do some research or check back through the book to help you decide what you think. Or talk about it with a friend.

GLOSSARY

atmosphere mix of gases surrounding the Earth up to the edge of space.

barrage human-made barrier across a water course, such as an estuary.

buoy floating device, usually anchored to the sea floor to prevent loss or damage.

climate normal pattern of weather over long periods.

condenser a piece of equipment that reduces gases to their liquid or solid form.

current flow of water or electricity.

displace cause to move usually as a result of conflict or environmental change.

downstream in the same direction as a river or stream current.

estuary wide, lower part of a river where it meets the sea.

evaporator a piece of equipment that turns a liquid into a gas.

fossil fuel fuel, such as coal, formed over millions of years from the remains of living things.

generator a machine that produces electrical power.

greenhouse gas gas, such as carbon dioxide, that stores heat in the atmosphere.

grid system of wires and pylons for sending electricity across a wide area.

head of water quanitity of water.

hydroelectric power using moving fresh water in rivers or from reservoirs to generate electricity.

infrastructure cables, roads, power stations and other structures needed to make a system work.

irrigation supplying farmland with water diverted from rivers or reservoirs using ditches, pumps, pipes and sprinklers.

kinetic energy energy produced by movement.

mains electricity electricity supplied through the grid to users from power stations.

microgeneration small-scale production of electricity to meet the needs of users.

non-renewable energy resource, such as coal that is running out as it is not replaced when used.

penstock angled channel or pipe taking water to a turbine for hydroelectric power.

renewable energy resource, such as water power, that is replaced naturally and can be used without running out.

reservoir human-made freshwater lake, usually formed by damming a river.

sluice gates gates used to control the flow of water in a channel.

subsidise pay to support something and encourage its success.

tide daily rise and fall of sea level.

INDEX

atmosphere 5, 10

barrage 16, 17, 21
buoys 19

climate change 5, 24, 25

dams 8–14, 16, 25, 28, 30
displaced people 12

energy 4–7, 17–20, 23, 24, 26–29

flood 12, 13, 20
fossil fuels 4, 5, 25, 30

generating electricity 4, 5, 7–9, 11,
    14, 15, 17–19, 22, 23, 26–29
greenhouse gases 5, 10, 24, 28
grid 14, 23

head of water 6
hydroelectric power 5–10, 12, 14–15, 22–23, 24,
    26, 28, 30

microgeneration 14, 15, 28

non-renewable resources 4, 24, 30

ocean thermal energy conversion (OTEC) 27

penstock 8, 14, 15
polluting 21, 30

renewable resources 5, 20, 24, 25, 28, 30
reservoirs 5, 6, 8, 10–13, 30

tidal current turbines 28
tidal power 5, 16, 17, 20, 21, 26, 28, 29
tides 5, 16, 17, 20, 29, 30
turbines 6–9, 14, 15, 17, 21, 24, 26–28

wave power 5, 18–21, 24, 26, 29